Solids in My World

Joanne Randolph

JOURNEYS®

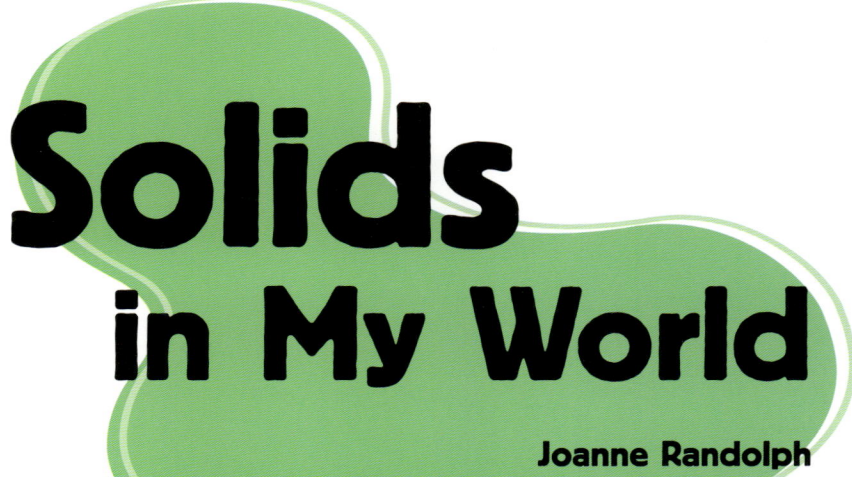

Rosen Classroom
New York

For Linda Lou and Lucas

Published in 2007 by The Rosen Publishing Group, Inc.
29 East 21st Street, New York, NY 10010

Copyright © 2007 by The Rosen Publishing Group, Inc.

All rights reserved. No part of this book may be reproduced in any form without permission in writing from the publisher, except by a reviewer.

First Edition

Photo Credits: Cover and p. 5 (rock), p. 5 (sky) © DigitalVision; Cover (yarn) © Getty Images; Cover and p. 7 (fire truck) Maura B. McConnell; pp. 5 and 22 (milk) © Digital Stock; p. 7 (acorn) © William Whitehurst/Corbis; p. 7 (skyscraper) © Michael Pasdzior/Getty Images; p. 9 © Rita Mass/Getty Images; p. 11 © Angelo Cavalli/Getty Images; p. 15 © Tom Stewart/Corbis; p. 17 © Nicki Pardo/Getty Images; p. 19 © Royalty-Free/Corbis; p. 21 © Alyson Aliano/Getty Images; p. 22 (frame), (pedal) © PhotoDisc; p. 22 (scarf) © Michael Keller/Corbis.

ISBN: 978-1-4042-8422-7
6-pack ISBN: 978-1-4042-9193-5

Manufactured in the United States of America

Contents

1 Sorting 4

2 What Is a Solid? 6

3 Solids Everywhere 14

4 Words to Know 22

5 Books and Web Sites 23

6 Index 24

7 Word Count 24

8 Note 24

Science is all about getting to know the world around us. We can start by sorting things into groups. We can sort things in our world into solids, liquids, or gases.

We are going to learn about solids. A solid has a size. Some solids are small. Sand and acorns are small solids. Some solids are big. A skyscraper and a fire truck are big solids.

Solids have a shape. A cup is a solid. A solid can change its shape if a force acts on it. If we break the cup, it is no longer shaped like a cup. Liquids and gases have no shape.

Solids can be hard or soft. A desk is a hard solid. A telephone is a hard solid, too. A blanket is a soft solid. Yarn is a soft solid, too.

Solids can feel rough or smooth. A brick is a rough solid. It feels bumpy when you touch it. A piece of paper is a smooth solid. It feels flat or even.

A bike is a solid. It is a solid that is made up of many solids. The wheels, the frame, the pedals, and the chain are all solids that make up the bike.

The clothes you wear are solids. Your hat, scarf, and mittens are solids. Your pants, socks, and shoes are solids, too.

Sometimes solids, liquids, and gases can change from one form to another. Ice is a solid. If the air becomes warm, ice melts. When the ice melts it turns to water. Water is a liquid.

Solids are everywhere in our world. What are some of the solids you see in this picture? Can you find the boat? The trees are solids, too. Can you name all the solids here?

Words to Know

frame

liquid

pedal

scarf

Here are more books to read about solids:
Solids, Liquids, Gases (Simply Science)
by Charnan Simon
Compass Point Books

Web Sites:
Due to the changing nature of Internet links, Journeys has developed an online list of Web sites related to the subject of this book. This site is updated regularly. Please use this link to access the list: www.powerkidslinks.com/mws/solids

Index

B
bike, 14

C
clothes, 16

G
gases, 4, 8, 18

I
ice, 18

L
liquids, 4, 8, 18

S
shape, 8

size, 6

W
water, 18

Word Count: 284

Note to Librarians, Teachers, and Parents

Journeys books are specially designed to help emergent and beginning readers and English language learners build their skills in reading for information. Sentences are short and simple, employing a basic vocabulary of sight words, simple vocabulary, and basic concepts, as well as new words that describe objects or processes that relate to the topic. Large type, clean design, and stunning photographs corresponding directly to the text all help children to decipher meaning. Features such as a contents page, picture glossary, and index introduce children to the basic elements of a book, which they will encounter in their future reading experiences.